MY FIRST SCIENCE BIOGRAPHY

Eugenie Clark

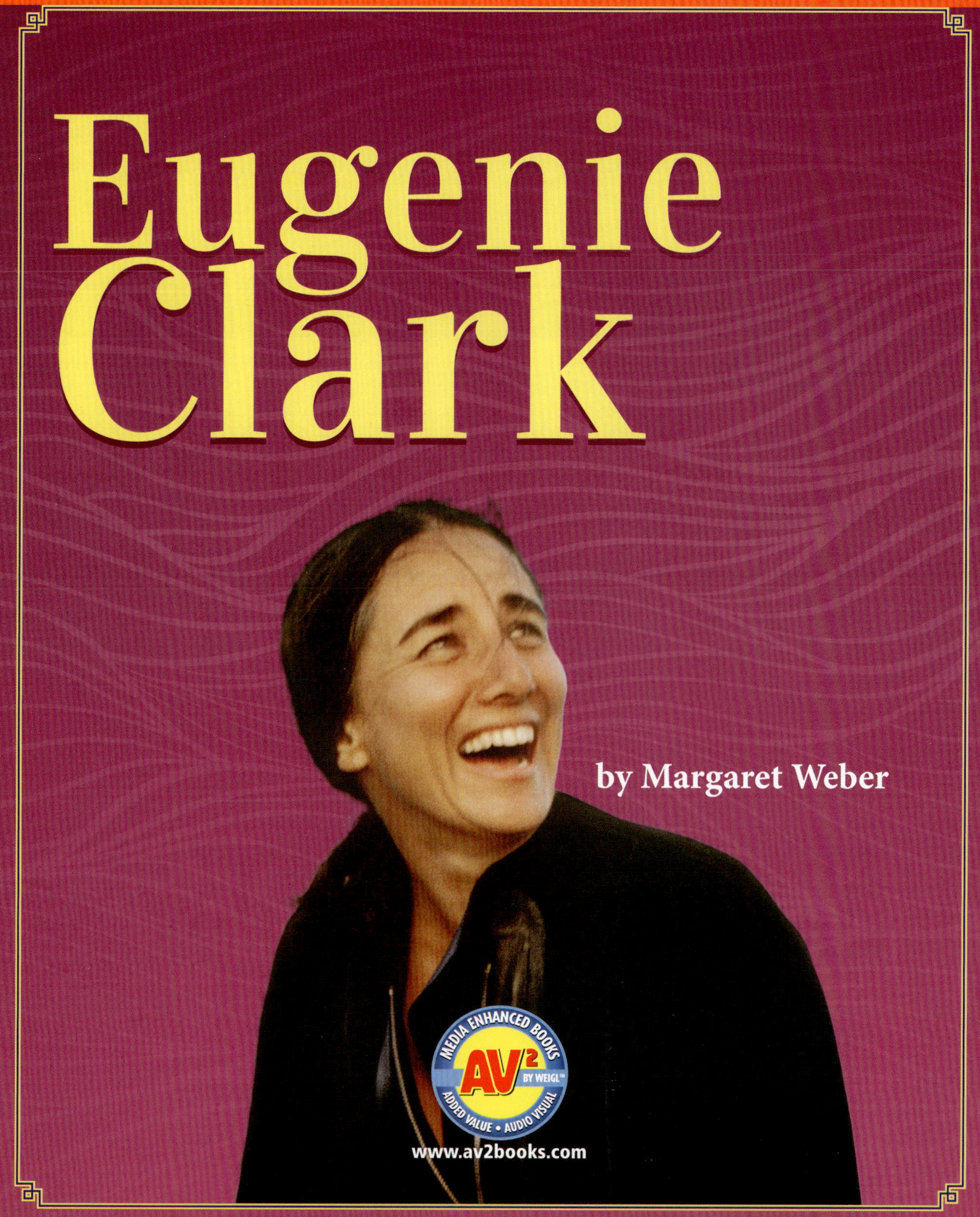

by Margaret Weber

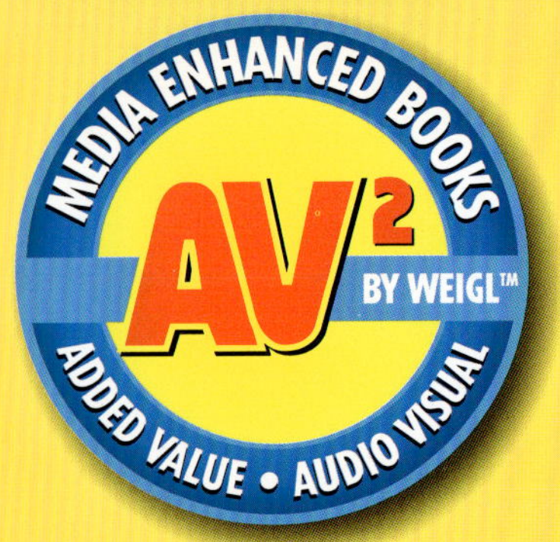

Go to www.av2books.com, and enter this book's unique code.

BOOK CODE

AVG76779

AV² by Weigl brings you media enhanced books that support active learning.

AV² provides enriched content that supplements and complements this book. Weigl's AV² books strive to create inspired learning and engage young minds in a total learning experience.

Your AV² Media Enhanced books come alive with...

Audio
Listen to sections of the book read aloud.

Video
Watch informative video clips.

Embedded Weblinks
Gain additional information for research.

Try This!
Complete activities and hands-on experiments.

Key Words
Study vocabulary, and complete a matching word activity.

Quizzes
Test your knowledge.

Slideshow
View images and captions, and prepare a presentation.

... and much, much more!

Published by AV² by Weigl
350 5th Avenue, 59th Floor
New York, NY 10118
Website: www.av2books.com

Library of Congress Control Number: 2019938600

ISBN 978-1-7911-1134-2 (hardcover)
ISBN 978-1-7911-1135-9 (softcover)
ISBN 978-1-7911-1136-6 (multi-user eBook)
ISBN 978-1-7911-1137-3 (single-user eBook)

Printed in Guangzhou, China
1 2 3 4 5 6 7 8 9 0 23 22 21 20 19

062019
311018

Project Coordinator: Heather Kissock
Designer: Terry Paulhus

Every reasonable effort has been made to trace ownership and to obtain permission to reprint copyright material. The publishers would be pleased to have any errors or omissions brought to their attention so that they may be corrected in subsequent printings.

Weigl acknowledges Getty Images, iStock, Shutterstock, Newscom, and Alamy as its primary image suppliers for this title. Page 11 BL: Smithsonian Institution Archives (SIA-SIA2008-199).

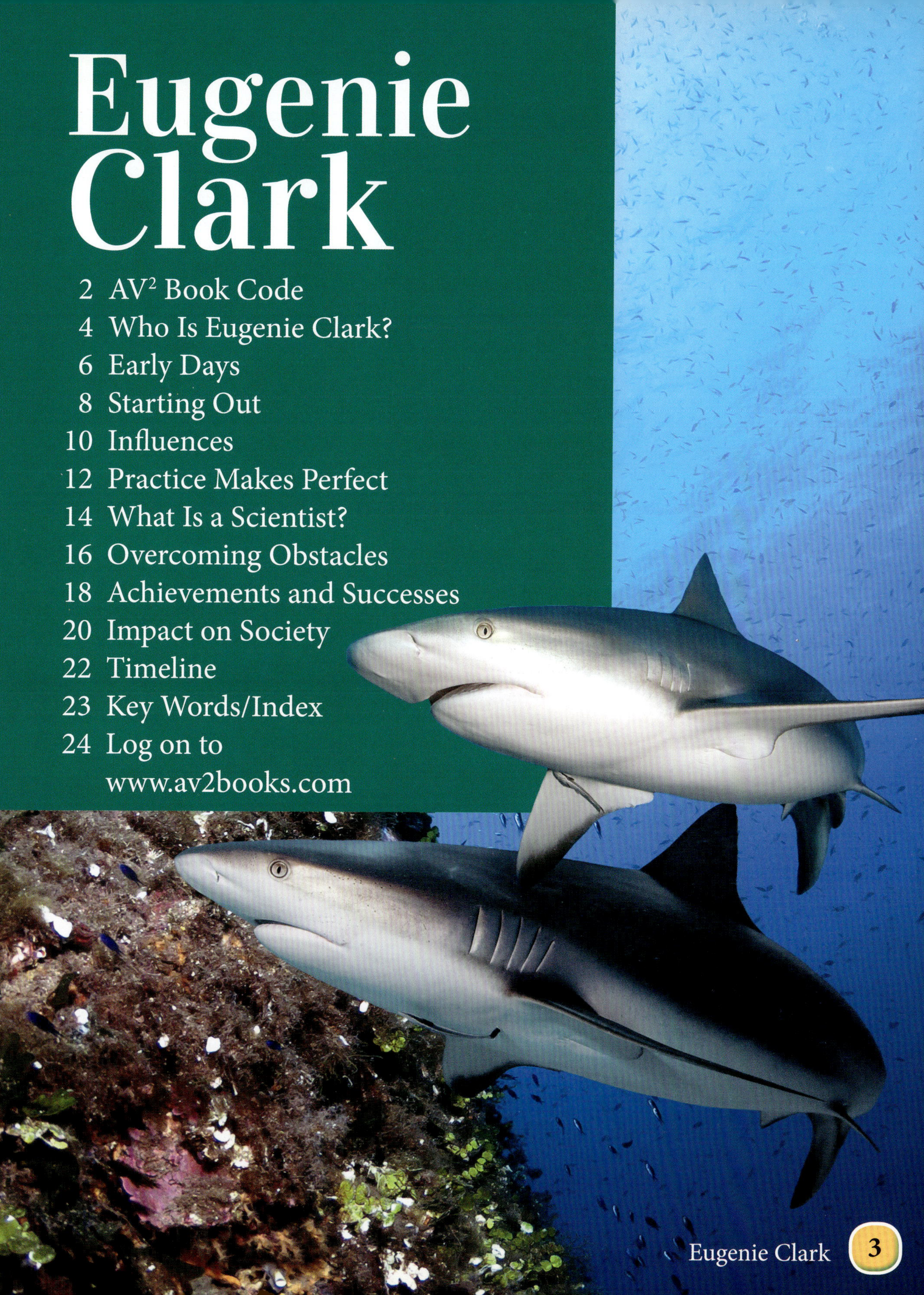

Eugenie Clark

Who Is Eugenie Clark?

Eugenie Clark is known as "The Shark Lady." She was a scientist. Eugenie studied **marine** life. She wanted to learn all that she could about sharks. Eugenie used what she learned to teach others about these animals.

Eugenie was one of the first scientists to use **scuba** diving to study life underwater. She also spent time fighting to keep oceans safe for wildlife.

"Sharks are among the most perfectly constructed creatures in nature. Some forms have survived for 200 million years."

Eugenie Clark worked to show that sharks deserved to be studied and were important to ocean life.

Early Days

Eugenie was born in New York City, New York, in 1922. Her father was American. Her mother's family came from Japan. Eugenie's father died when she was 2 years old. Her mother had to raise Eugenie on her own.

When her mother went to work, Eugenie would often go to the New York **Aquarium**. It was home to many kinds of sea life. The aquarium was where Eugenie discovered her love of the ocean.

The New York Aquarium is located on Coney Island. It is home to 350 different types of sea life.

Where Is New York City?

New York City is found in the northeast part of the United States. About 8.6 million people live in New York City. More people live there than any other city in the country.

Today, scientists who want to study sharks often use a shark cage. It is a structure that protects divers from sharks while they are in the water.

Starting Out

In 1938, Eugenie went to Hunter College in New York City. There, she studied **zoology**. She learned how to study animals. Eugenie earned her **Ph.D.** at New York University (NYU). During this time, she also worked as a research assistant in California. Research assistants help scientists with their work.

While in California, Eugenie learned how to dive underwater. She used early forms of scuba gear. This equipment let her breathe underwater. In 1951, she studied sharks in the Red Sea. She was one of the first scientists to do so.

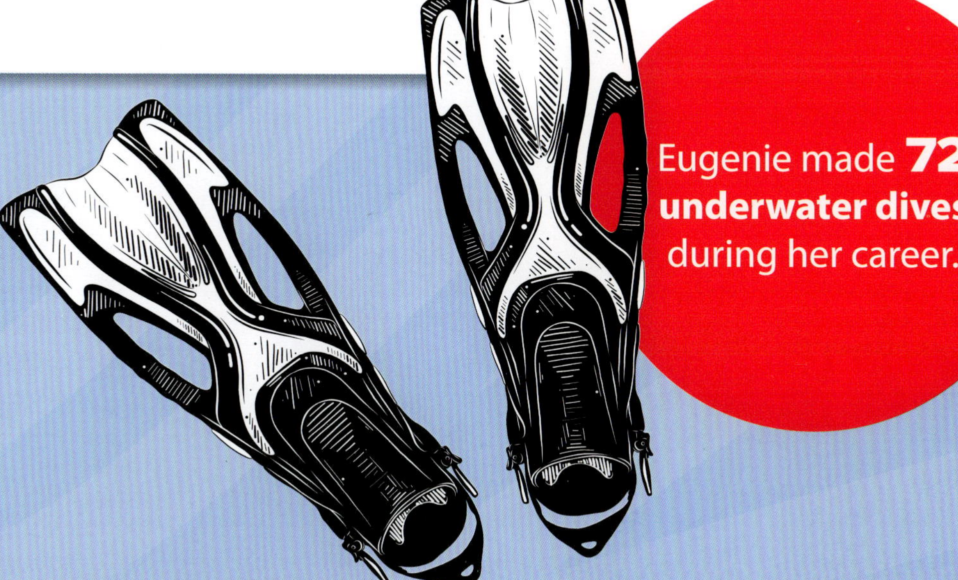

Eugenie made **72 underwater dives** during her career.

Influences

Eugenie was inspired by many people. Her teachers encouraged her. Her scientist friends showed her how to gather information. Some people even gave Eugenie money to help with her work. Over time, Eugenie became "The Shark Lady."

Eugenie was comfortable swimming with sharks because she had studied them for years.

Dr. Charles Breder, Jr.

Dr. Charles Breder, Jr. was one of Eugenie's teachers at NYU and a lifelong **mentor**. Charles supported her work with sharks.

Dr. Myron Gordon

While at NYU, Eugenie worked with Dr. Myron Gordon. She was his research assistant.

Anne and William H. Vanderbilt

Anne and William H. Vanderbilt were **philanthropists**. They gave money to help build the Mote Laboratory and Aquarium. This was where Eugenie did her research.

Practice Makes Perfect

When Eugenie was young, there were not many women scientists. Some people were surprised that she wanted to study sharks. She had to prove that she could do the work.

People thought of sharks as mean animals that were not smart. Eugenie wanted to show people that they were wrong. She trained sharks to press a button to get food. This showed that a shark could learn. Her discoveries changed the way people all over the world thought about sharks.

Eugenie was the first person to discover that sharks do not need to keep **moving** in order to breathe.

In 1995, Eugenie and her team discovered that whale sharks give birth to live babies. They do not lay eggs.

What Is a Scientist?

A scientist is someone who studies things. People who study animals are called zoologists. Scientists are very curious. They love solving problems. Scientists try to answer questions through **experiments**. They collect information using a six-step method.

Eugenie had to learn about the ocean to understand sharks more. She used tools to measure water temperatures and how fast the water was moving.

Using a 6-Step Scientific Method

STEP 1

QUESTION

Scientists ask a question about what they want to learn. They read books and go online to research what other people know about the topic.

STEP 2

HYPOTHESIZE

The scientists then guess what the answer to their question might be. This guess is called a hypothesis.

STEP 3

EXPERIMENT

Scientists plan an experiment to see if their hypothesis is right. They gather the materials they need. Then, they set the materials up and do the experiment.

STEP 4

OBSERVE & RECORD

Scientists use their senses to observe what happens during their experiment. They watch. They smell. They listen, touch, and even taste. They then record what they find.

STEP 5

ANALYZE

Scientists think about what happened during their experiment. They decide if the experiment showed that the hypothesis was right or wrong.

STEP 6

SHARE RESULTS

Scientists let other people know about their experiment and what they found out. They may write a report or give a speech.

One obstacle Eugenie did not have to overcome was fear. Many other scientists were afraid of sharks, but Eugenie was not.

Overcoming Obstacles

Eugenie had four children. It was often difficult to balance her family and her work. She took her children with her on research trips. She also taught them to swim and dive at young ages. This allowed her family to be a part of her work.

Eugenie was just beginning her scientific work when World War II (1939–1945) began. Some people did not trust Eugenie because she was Japanese-American. Japan and the United States were at war. Eugenie had to show that she could be trusted. She did this by working hard and staying focused on science.

Eugenie sometimes worked together with other scientists. She published scientific papers with fellow researcher, Professor Adam Ben-Tuvia.

Achievements and Successes

Eugenie studied more than sharks. She also studied fish. She even discovered new kinds of fish. In the Red Sea, she found a fish that made its own shark **repellent**. It is called the Moses sole.

Eugenie won many awards. She was given a New Orleans Grade Isle (NOGI) award. It is for people who are important to the diving community. She won for the important research she did while diving.

Eugenie helped to create the Mote Marine Laboratory and Aquarium in Sarasota, Florida. At first, it was a small lab. It focused only on sharks. Today, it has expanded to include all marine research.

Eugenie named a fish she discovered after her son. The Trichonotus nikii is named for her son Nikolas.

Although the Mote Marine Laboratory started out studying sharks, researchers there now study manatees, dolphins, sea turtles, and coral reefs as well.

Impact on Society

Eugenie inspired people to study marine life. Many women see her as a role model. She inspired many young women to become scientists.

Eugenie left a lasting impact on marine science. She gave talks on sharks around the world. Her discoveries taught scientists new things about ocean life. Her research still helps scientists today. The Mote Marine Laboratory continues her fight. Its scientists work to keep the oceans clean and healthy.

Eugenie taught at the University of Maryland. She shared her knowledge with students and encouraged them to complete their own research.

Timeline

1922
Eugenie is born in New York City.

1951
Eugenie makes her first **expedition** to the Red Sea.

1953
Lady with a Spear, Eugenie's first book about her life, is published.

1955
Eugenie opens the Cape Haze Marine Laboratory. This eventually becomes the Mote Marine Laboratory.

1999
Eugenie retires from her professorship at the University of Maryland.

2014
Eugenie turns 92 years old and celebrates with a dive in the Gulf of Aqaba in the Red Sea. This is her final dive. She dies the following year.

Key Words

aquarium: a building that is open to the public and features large glass tanks filled with sea life

expedition: a journey that is taken for exploration or to complete scientific research

experiments: tests that are done to prove a hypothesis

marine: found in the sea

mentor: someone who gives advice and helps someone else achieve his or her goals

Ph.D.: a title that shows a person has received the highest level of education possible in a topic

philanthropists: people who donate money to a cause that is important to them

repellent: a substance that keeps pests from coming near

scuba: a device that allows divers to breathe underwater

zoology: the study of animals

Index

Log on to www.av2books.com

AV² by Weigl brings you media enhanced books that support active learning. Go to www.av2books.com, and enter the special code found on page 2 of this book. You will gain access to enriched and enhanced content that supplements and complements this book. Content includes video, audio, weblinks, quizzes, a slideshow, and activities.

AV² Online Navigation

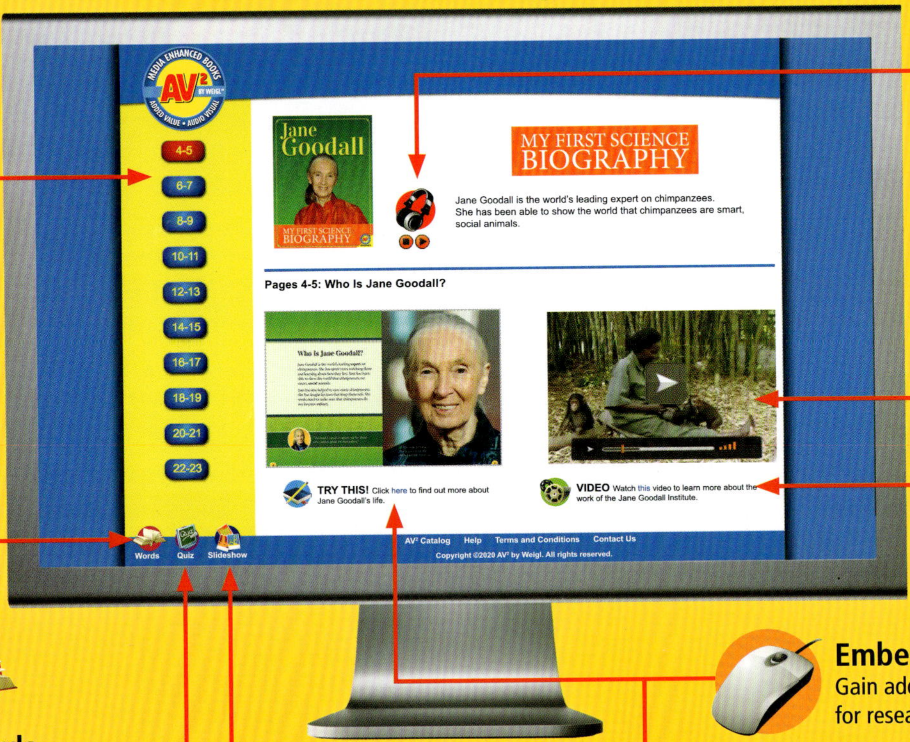

Audio
Listen to sections of the book read aloud.

Book Pages
AV² pages directly correspond to pages in the book.

Video
Watch informative video clips.

Embedded Weblinks
Gain additional information for research.

Key Words
Study vocabulary, and complete a matching word activity.

Try This!
Complete activities and hands-on experiments.

Quizzes
Test your knowledge.

Slideshow
View images and captions, and prepare a presentation.

AV² was built to bridge the gap between print and digital. We encourage you to tell us what you like and what you want to see in the future.

Sign up to be an AV² Ambassador at www.av2books.com/ambassador.

Due to the dynamic nature of the internet, some of the URLs and activities provided as part of AV² by Weigl may have changed or ceased to exist. AV² by Weigl accepts no responsibility for any such changes. All media enhanced books are regularly monitored to update addresses and sites in a timely manner. Contact AV² by Weigl at 1-866-649-3445 or av2books@weigl.com with any questions, comments, or feedback.